REDUCE YOUR CARBON FOOTPRINT

A BEGINNERS GUIDE TO REDUCING YOUR GREENHOUSE GAS EMISSIONS

ANN M COLSON

Disclaimer

TABLE OF CONTENTS

INTRODUCTION

So, you've seen the words 'Reduce your Carbon Footprint' sprayed onto a wall in the form of graffiti whilst in the midst of a conversation that seemed to matter more, and likely did not pay too much attention to it, right? Well, you're not alone. There are millions of people out there who do not pay much attention to the fact that there are lots of little things that they can do as an individual to change the overall 'health' of the world for the better.

Our carbon footprint is increasing at an alarming rate. The rising temperatures and shifting precipitation patterns are having a profound effect on the growth pattern of plants, as well as the sea levels which are rising. This is all thanks to our 'carbon footprint'. That's all really rather easy to understand, isn't it?

If only it were as easy to realize that as the torchbearers of our society, we have a greater role to play in ensuring the survival of our environment. It is after all, the very thing that has nurtured and will continue to nurture all of us selflessly until – well, until the day it is no more! This is a truth that sadly will come to fruition if we do not start to behave far more responsibly. This book will not only teach you how to calculate what your 'carbon footprint' actually is, but will also show you the ways in which you can take steps to reduce it on a personal level, and therefore help to prevent all sorts of terrible things happening to this beloved planet of yours.

If you're a non-vegetarian, you've probably already been confronted with the idea that if you gave up eating meat, you would be able to reduce the number of animals mindlessly slaughtered. You have however probably rubbished that notion several times over, thinking 'How could one person possibly make a difference?' Deep down however you know very well that if there were several others like you, it would be possible to significantly reduce the number of animals slaughtered.

The same can be said about reducing your carbon footprint. There are of course things that need to be done on a larger scale; which will also be discussed during the course of this book, but there are so many things that people like you and I can do to prevent the destruction of entire towns and cities due to global warming, and the resulting flooding.

You wouldn't throw a paper cup on the road would you? That would mean littering your environment, right? Well, then why wouldn't you take some simple steps towards ensuring that the very environment you love is protected from destruction? By reading this book you will learn everything you ever wanted to know about your 'carbon footprint', including how you can do your bit to reduce it instead of continuing in your current footprints which will ultimately lead us all to a terrible destination; much like the Pied Piper did with the innocent children of Hamelin.

CHAPTER 1: WHAT 'CARBON FOOTPRINT' MEANS

In the introduction, we have briefly shown that we can have some degree of control over our carbon footprint, but do we really understand what it means? The purpose of this chapter is to throw some light on exactly what a carbon footprint is, and why it matters in the grand scheme of things; or to put it more accurately; why the reduction of it matters in order to sustain a healthy environment.

- What is a carbon footprint?

The definition is:

'The total amount of greenhouse gases produced to directly and indirectly support human activities'; it is usually expressed in equivalent tons of carbon dioxide (CO2).

Still don't get it? Well, let's look at it in layman's terms. Think of it in relation to driving a car. The engine burns the fuel which as a byproduct produces carbon dioxide, the amount produced depends on the fuel consumption of the car and the distance travelled. This same carbon dioxide is produced when you heat your house with fuels such as oil, gas and coal, and even if you use electricity to heat your house, CO2 is still produced as part of the process which generates this electricity. Even when you go to the market to buy foods and other goods, you must realize that those very foods and goods have been produced using processes that involve the creation and therefore emission of carbon dioxide.

When we talk about 'your' carbon footprint, this is the sum of all the emissions of carbon dioxide; i.e. 'Greenhouse gases' that were produced as a result of your activities in any given time period. The normal timeframe used when calculating someone's carbon footprint is one year. Here are some examples of the amounts of carbon dioxide that are produced as a result of everyday activities/processes that we take for granted; simple processes that are doing significantly more harm to the environment than we can possibly imagine.

- Every gallon of gasoline fuel (US) used in engines; leads to the emission of 8.7 kg of CO2.

- Every gallon of oil (US) used for heating; leads to the emission of 11.26 kg of CO2

- Every 10 – 12km travelled by train or bus; leads to the emission of 1 kg of CO2.

- Every 2.2 km travelled in an airplane; leads to the emission of 1 kg of CO2.

- Operating your computer for 32 hours; leads to the emission of 1kg of CO2.

- The production of 5 plastic bags; leads to the emission of 1kg of CO2.

- The production of 2 plastic bottles; leads to the emission of 1kg of CO2.

- The production of 1 cheeseburger; leads to the emission of 3.1 kg of CO_2.

Now that you have a clearer understanding of what is meant by 'greenhouse gases', you know it has nothing to do with actual 'greenhouses'. Of course, there are other 'greenhouse gases' in addition to CO_2 that are emitted as a result of your activities; such as methane and ozone. These are also taken into account when calculating your carbon footprint; this is done by converting them into their equivalent amount of 'carbon dioxide'. It is important to appreciate that they, therefore, have the same effect as CO_2 on the phenomenon known as 'global warming'. Now who would have known that?

Some people take the process of monitoring their carbon footprint very seriously, and can in fact quote a definitive figure; this is usually expressed as an amount of CO_2 or sometimes even as an amount of carbon.

It is possible to calculate the actual amount of carbon produced by multiplying the kg's of CO_2 by (.27). This is because 1000 kg of CO_2 is equal to 270 kg's of carbon.

It's essential that you understand how important it is to reduce your carbon footprint. We all know that smoking is seriously bad for your health, and the very real danger of cancer remains the number one reason that people quit.

In the next chapter we will be looking in more detail at the danger of us not taking responsibility for dramatically reducing our own carbon footprint; we must all do our bit in order to protect our extraordinary planet.

Chapter 2: Why a Carbon Footprint Matters

We all have a sense of morality from childhood however perhaps this sense of morality has its limits. We will for instance often do bad things simply because we are unaware of the consequences that our actions will have either on us or our environment.

Would you change your behavior if one day you were told that even the little things you do on a daily basis actually have an enormous negative impact on the environment?

Why Carbon Footprints Matter

It is essential that we understand the impact we have on our planet; after all it does sustainment our lives, and the lives of all other species of animal and bird. For this reason, we need to take all the necessary steps to reduce our carbon footprints. There are of course a number of other obligations that we have to our environment, but this is certainly amongst one of the most important.

Our careless attitude towards global warming threatens multiple species of birds and animals; for instance, what happens when the vegetation patterns shift and the migratory birds and animals that arrive after a long and arduous trek, find that their food isn't there; having either bloomed too early or not at all? Additionally; what happens to the Arctic polar bears when they lose their hunting ground due to melting ice? These are just a couple of examples of what might lead to either the extinction of several species of animals and birds, or serious changes in their behavior which could then adversely affect our lives.

The impact of global warming on humans goes far beyond the 'flooding' of our towns and cities. Shifting crop patterns could for instance result in food shortages; children are already dying of hunger every single day in the world, just imagine how many more would die if drought became a worldwide issue.

There are also diseases like 'malaria' that are currently on the increase simply because the warmer climate is more favorable to mosquitoes.

Additionally the rise in air pollution has resulted in a significant increase in respiratory diseases; asthma and certain other allergies are more prevalent now than they have ever been before.

There are several economies that are more dependent on land and natural resources than others, and as a result these are already being significantly negatively affected as a result of our inability to control our carbon footprint. For example; the increase in sea temperatures is threatening to destroy the coral reef which on its own is a multimillion-dollar business.

Of course changing our entire lifestyles in order to reduce our carbon footprints is easier said than done, however through education we truly can pave the way towards a better future. By taking the reins and reducing our carbon footprint, we will make massive strides towards both protecting and preserving our planet for future generations.

To ensure success it is imperative that we all remember; we are not alone. Looking at the vegetarian versus non-vegetarian comparison again; there are indeed many people out there currently converting to a vegetarian diet, the same is also true of people becoming more environmentally aware.

A number of people will be able to imagine jumping on the bandwagon of reducing their carbon footprint; however others will still feel that their efforts will be wasted. Rest assured this is not the case; every single person can make a difference. Don't wait, get on that bandwagon now!

Chapter 3: The Silent Culprits of Carbon Footprints

Often unbeknownst to us there are multiple process and behaviors occurring as part of our daily lives that are contributing to the size of our carbon footprint.

In order for us to start measuring our own carbon footprint and therefore taking the necessary steps towards reducing it, it is important that we not only understand the activities that ultimately result in the emission of greenhouse gases, but also how significant each activities impact is, so that we can target the biggest culprits and therefore make the greatest impact.

- **Travelling**

We have already seen in the first chapter that 8.7 kg of carbon dioxide is emitted every time a gallon of gasoline is used. If you consider how often you drive your car, it is obvious that you use significantly more than a gallon of fuel and therefore in turn, obvious that you produce significantly more than 8.7kg of CO2; all as a result of the seemingly 'harmless' act of driving.

The use of public transport in the form of buses and trains is less damaging than driving yourself, however it still has an impact; for every 10 - 12km travelled you are contributing to the emission of 1kg of CO2. Air travel is even more damaging; with 1kg of CO2 being produce for every 2.2km travelled and if you consider the average distance of a flight, this equates to a huge emission of CO2.

In today's world we have become accustomed to being lazy; gone are the days when people would walk to the supermarket. People find the need to drive, or to be driven for even the shortest of journeys in today's world. This is resulting in a silent increase in their carbon footprint on almost a daily basis.

- **The use of appliances**

Who can do without the use of appliances in today's world? We saw in the first chapter that a kilogram of carbon dioxide is emitted when we use a computer for 32 hours, however as we all know that is not the only appliance we are using on a daily basis these days. There are several other appliances such as electric dryers; refrigerators; freezers; and dishwashers, all of which emit carbon dioxide every single time they are used. We have become so dependent on appliances that it is frightening to think of a world without them. Just imagine the number of appliances per household, and then the number of households; you can then start to understand how high the amounts of carbon dioxide being released into the air are, and therefore how serious the looming dangers are.

- **Water**

Who can live without water? Well the truth is, nobody. However the fact of the matter is a lot of people are using water excessively and for non-essential activities.

The waste of water has become a very serious issue and does nothing to help reduce our carbon footprint. Even the 'essential' uses of water are a concern; for example in regards to drinking water, the very action of cleaning the water to ensure it is safe to drink involves the use of a great deal of energy, the consumption of this energy results in the emission of CO2.

- **Electricity and heat generation**

What's a world without light? Of course we need electricity in our day to day lives however do we all understand the impact it has on our carbon footprint?

Almost all the industrialized nations get the majority of their electricity from the combustion of fossilized fuels. Consequently you will find that the energy supplied by your energy provider has a huge role to play in the size of your carbon footprint.

The use of electricity by major industries is of course significantly higher when compared with others like residential. This is because of the energy-intensive processes they use, it is however really important to realize that due to their reliance on electricity, even the residential and commercial sectors use large amounts, especially when it comes to things like lighting, air-conditioning, appliances and heating.

- **Industrial**

Of course in the above point we have seen that the consumption of electricity is at an all-time high as far as the industrial sector is concerned. Aside from the electricity used by the industrial sector, because their manufacturing and industrial processes also all produce large amounts of greenhouse gases; specifically a lot of carbon dioxide, they are one of the biggest culprits when it comes to the size of their carbon footprint. This is mainly because many of these manufacturing processes involve the burning of fossilized fuels in order to produce the heat and steam that are integral to these processes.

- **Land use changes**

This happens when humans convert land from its natural state into something that can be used for things such agriculture and/or building settlements.
It is one of the glaring examples of how we are destroying the environment.

The worse example of this is the process of deforestation. Deforestation is the removal of the trees that have lived for years in our forests. These forests are being wiped out either for the timber they can supply or simply so that the land can be converted into farmland.

This process results in high levels of greenhouse gases being released into the atmosphere, however that isn't it; what many people don't know is that trees actually provide a very crucial role; it is their job to clean the atmosphere of excess CO2 through the process of photosynthesis. Their removal therefore has a long term impact on carbon dioxide levels.

Furthermore, this deforested land becomes vulnerable to the elements and the rate of soil erosion increases, which further hampers the ability of the soil to act as a 'carbon cleanser' and as a result CO2 levels continue to rise.

- **Food**

The production of food impacts on the levels of CO2 in a number of different ways; not only due to the changes in the use of the land, talked about above; but also because of the natural emissions that come from livestock such as cows and the production of certain food types such as rice.

There are also other factors that come into play, such as the transportation, storing, cooking and disposing of the food that we eat. These all play a significant role in the level of carbon dioxide emissions produced.

In the US, each household produces 48 tons of greenhouse gases and surprisingly after transport and housing; food is one of the largest contributors towards the size of your carbon footprint.

- **Nuclear energy**

A lot of lobbyists from the nuclear energy industry will argue that the production of energy via nuclear power stations does not produce carbon dioxide. However that is fundamentally untrue. Carbon dioxide is produced as part of the production of atomic energy, albeit at a smaller level. In fact, even the production of energy from renewable sources like solar, water and wind are responsible for the emission of carbon dioxide to some extent at least.

In short, whenever there is an industry that is involved in the manufacture of electricity, there will also be levels of carbon dioxide being pumped into the atmosphere.

- **Natural sources**

Besides the aforementioned 'human' sources of carbon dioxide emissions, it is important to note that Mother Nature herself is continually contributing towards the level of CO2 in the atmosphere. The human sources of carbon dioxide are really bad, however because they upset the 'natural balance' in the carbon cycle that existed before the Industrial Revolution, the planet reactions to this imbalance by pushing even more CO2 into the air. Most of the naturally produced carbon dioxide comes from the ocean-atmosphere exchange, as well as other sources such as plant, animal and soil respiration.

We have seen in this chapter that humans are primarily accountable for creating the huge carbon footprint that is responsible for the dangers associated with global warming.

You might now however be thinking; but we can't live without things like electricity, food and water; and you're right, however you need to realize that there are still ways to reduce your own carbon footprint dramatically by being smarter and more efficient. These will be discussed in the course of this book.

Most of us live in denial about how much we personally impact on global warming, and we also feel that there is nothing that we can now do to reverse it. The purpose of this book however is to show you that you can most definitely make a difference.

Before I go into more detail about how you can start reducing our carbon footprint, it is important that you learn how to measure your own footprint and therefore fully appreciate just how responsible you personally are for the ever increasing levels of greenhouse gases in the atmosphere.

The following chapter will endeavor to do just that. Once you are absolutely clear on what our carbon footprint is, and what processes/behaviors in our life are responsible for the greatest production of CO_2, we will move on to learning how to reduce it.

CHAPTER 4: MEASURING YOUR CARBON FOOTPRINT

So you've learned about all the potential dangers resulting from our ever increasing carbon footprints, well now we need to learn how to measure our own footprint?

Let us start by learning and understanding the methodology used to measure it effectively, and from there we can then start taking the necessary steps to reduce it.

You need to start by calculating your share of the energy used in the home, as well as the amount of water used and any waste disposal.

In all probability you are not the only person living in your home so in order to arrive at your individual share you have to calculate the total energy, water and waste disposal values, and divide them by the number of people living in your home.

- **Electricity**

For most people the electricity that is used comes from a grid that is rather complicated, as a result you may therefore wonder how you can estimate the amount of carbon dioxide that is emitted, don't worry about that.

There are published 'direct emissions' factors that show the amount of emissions produced by power stations in order to produce an average kilowatt-hour within a particular region. In addition to this there are also details about 'indirect emissions' and 'transmission losses', and when these are combined with the 'direct emissions' figure, a realistic picture of the amount of energy that is ultimately used can be calculated.

The following is the formula used to calculate the 'electricity' quotient of that carbon footprint:

$$\text{Use (kWh/yr)} * \text{EF (kg } CO_2e\text{/kWh)} = \text{emissions (kg } CO_2e\text{/yr)}$$

- **Fuels**

There are several houses out there that use fossil fuels as a primary source of energy as opposed to electricity; things like natural gas, liquid petroleum (LPG) and coal. It is easy to calculate the direct emissions from the combustion of these fuels because they are the same throughout the world; this is because it is dependent on the physical properties of the concerned fuel.
The 'indirect emissions' from the fuel depends on the technology used to prepare and transport the fuel, along with the distance that it needs to travel. When you combine the 'direct' and the 'indirect' factors you get emission factors for the full life cycle of the fuel.

In the case of LPG usage the formula would be something like this:

$$\text{LPG: use (litres/yr)} * \text{EF (kg } CO_2e\text{/litre)} = \text{emissions (kg } CO_2e\text{/yr)}$$

- **Waste Disposal**

The emissions produced as a result of the process of waste disposal are mainly methane, this is the gas produced both at the landfill sites themselves and also as part of the transportation process. All you have to do is calculate the amount of waste that you produce every week, and then multiply it by 52; you will then have your annual waste production value. In order to get your footprint, you then multiply that figure by the carbon intensity:

Use (kg/week) * 52 * EF (kg CO_2e/kg) = emissions (kg CO_2e/yr)

- **Water Use**

As far as your water usage goes, you will be surprised at the amount of carbon that can be produced in this process. There are actually two processes that lead to these emissions; the electricity that is used to pump the water into households, and the amount of nitrous oxide produced as part of the wastewater and sewage treatment processes.

These values will differ from country to country but once you work out your daily usage and multiply it by 365, then you will have your annual usage. Apply that carbon intensity and you will arrive at your carbon footprint value as far as water usage is concerned.

Use (ltrs/day)* 365*EF (kg CO_2e/kWh) =emissions (kg CO_2e/yr)

- **Transport**

In order to calculate your travel footprint you have to estimate the amount of travel that you have done in the past 12 months; this must include all different modes of transport.

Personal vehicle:

Let's begin with your personal vehicle as this is by far the most common means of transport. For this you have to multiply the distance that you have driven in the last year, with the emissions factor for your vehicle. While it is easy to estimate the distance that you have driven in the past 12 months, it is not always simple to estimate the emissions factor for your vehicle.

You need to use an carbon intensity that includes 'direct emissions' from actual fuel use, 'indirect emissions' from fuel production, and an estimation of the emissions involved in the construction of the vehicle itself:

Distance (km/yr) /*EF (kg CO2e/km) = emissions (kg CO2e/yr)

Public Transport

The emission factors that are associated with public transport are often published by the government in the countries that they exist. These include both 'direct' and 'indirect' emissions. You can then add to these the vehicle construction emissions; these are typically much lower for public transport vehicles than for private vehicles.

Let us take the example of travelling by 'bus'; the formula designed to calculate that 'travel footprint' is:

Distance (km/yr) * EF (kg CO2e/km) = emissions (kg CO2e/y)

Flying

To calculate your 'flying' footprint you need to work out the distance that you have flown in the last 12 months. Once you have this distance you can follow the same method you did in the case of public transport, by using the emission factors that are published by government agencies. The emission factors vary for short, medium and long haul flights. It also makes a difference if you fly first class, business or economy. Of course you will also need to add in those construction emissions as well, these are calculated by passenger mile or passenger kilometer:

Distance (km/yr)* 1.09 * EF (kg CO_2e/km) = emissions (kg CO_2e/yr)

- **Food**

When it comes to calculating your 'food' footprint, you need to estimate the amount of food you have consumed and the emissions that were produced as a result of that food being supplied. You can use expenditure and weight to calculate the consumption of food however calculating expenditure isn't simple and is prone to errors, and trying to calculate the weight of everything eaten just isn't practical. Instead you can use daily food energy consumption figures (kcal/day).
Start by dividing your diet into the major food groups that have similar emission factors like: red meats; white meats; vegetables; cereals and dairy. You can use these food groups to get a sense of how your daily food intake adds up.

Normally a person needs 2000 to 3000 kcal/day. You can use IO-LCA and LCA literature in order to arrive at the emission factors for the foods involved. Once you have these emission figures you can begin the process of calculating those 'food footprint values', using the sample formulae below:

Red meat: cons. (kcal/day)*365*EF (kg CO2e/kcal) = emissions (kg CO2e/yr)

White meat: cons. (kcal/day)*365*EF (kg CO2e/kcal) =emissions (kg CO2e/yr)

Dairy foods: cons. (kcal/day)*365*EF (kg CO2e/kcal) = emissions (kg CO2e/yr)

- **Household**

Believe it or not, your household emissions can contribute significantly to your overall carbon footprint. This is also known as the 'products' footprint and involves calculating a summary of the footprints for all the products that you purchase.

You need to focus on expenditure in certain product groups and then use estimates from Input-Output Life Cycle Assessment (IO-LCA) in order to get the average emission factors for each group.

First we have to work out the average monthly expenditure in each of the product groups like: household; clothes; medical; recreational and any others. You then need to multiply these values by 12 in order to get the annual consumption for each group. Each annual consumption value then needs to be multiplied by an emissions factor

($kgCO_2e/\$$) which is calculated using averages from the IO-LCA literature. Once you have completed this for each and every one of the product groups, you will have a total products footprint.

Household: spend (\$/month) * 12 * EF (kg $CO_2e/\$$) = emissions (kg CO_2e/yr)

Clothes: spend (\$/month) * 12 * EF (kg $CO_2e/\$$) = emissions (kg CO_2e/yr)

Medical: spend (\$/month) * 12 * EF (kg $CO_2e/\$$) = emissions (kg CO_2e/yr)

You may now be thinking that the process of calculating your individual carbon footprint is too complicated, and I completely understand that it does appear a bit daunting; however there are a number of fantastic websites that make this whole process much easier by talking you through it, step by step.
This website can be used by people all over the world as it allows you to select your country of residence:

http://www.carbonfootprint.com/calculator.aspx

This website is targeted more specifically as the US market:

http://www.nature.org/greenliving/carboncalculator/

Why not have a go? You'll be amazed at how easy it actually is!

At the end of this process you will have a figure telling you the size of your total carbon footprint . Now that we have this, we can focus on what pivotal steps we need to take in order to reduce it. This will more often than not be achieved by reducing a number of our smaller footprints to achieve a cumulative effect.

Chapter 5: The 'Long' Steps to reducing Carbon Footprints

The road towards reducing our collective carbon footprint is a long and arduous one, and even though there are many things that we can do on an individual basis to ensure that we play our part in 'shrinking' the collective carbon footprint, there are of course things that really need to be done on a much 'larger' scale in order to reduce the risks associated with global warming.

Let us look at some of the endeavors that have been designed with the primary purpose of shrinking the global carbon footprint:

- **The Kyoto Protocol**

The Kyoto Protocol is an agreement that was negotiated in 1997 by many countries and finally came into force following Russia's ratification on February 16, 2005. The reason there was such a long delay between the agreement being initiated and finally set into motion was that it required a minimum of 55 countries to ratify it. In addition to this it was also necessary that the total emissions of those 55 countries made up at least 55 percent of the total global emissions of greenhouse gases. This protocol was developed under the UNFCCC – The United Nations Framework Convention on Climate Change.

What does the Kyoto protocol state?

It states that those who have ratified the Kyoto Protocol are committed to cutting the emissions of not only carbon dioxide, but of all other greenhouse gases as well, these include; methane, nitrous oxide and hydrofluorocarbons. If the countries that are participating have emissions that go beyond their targets, in order to offset this they are required to buy 'credit' from other participating countries that are well within their targets. The primary goal of the Kyoto Protocol is to ensure that the participating countries reduced their emissions of gases by 5.2 %, below the emission levels of 1990, by the year 2012.

This figure of 5.2 % is really a cumulative one as individual countries were assigned different targets based on their size. For the USA in fact, it was a target of 7 %. It is interesting to note that although India and China were two of the 55 countries to ratify the agreement, they were exempt from targets of any kind because they were deemed as 'developing countries' and as such, were not held culpable for the effects of global warming that began at the time of the Industrial Revolution.

Since 2005 a number of other countries have ratified the Kyoto agreement, in fact to date 169 countries have now ratified it. It is very interesting to note that the USA did not ratify the agreement until late in 2007 and Australia negotiated very hard during the process, stating that it wanted an eight percent increase in the emissions target that was initially allocated to them. The process of ratification is not one that is taken lightly as it is not only a contractual agreement of the highest order, but it also carries legal obligations. So while almost every country in the world might have signed the agreement, they also have a big responsibility to play in implementing it successfully.

Is the Kyoto Protocol a success or a failure?

It is obvious that it was implemented with the best of intentions, however when you look around today and see the fact that both carbon dioxide levels and global temperatures are continuing to increase at an alarming level, it has not perhaps been as successful as it was intended to be.

The problem when decisions are taken at such a high level is that they often involve a lot of politics. Many of the decisions that are made at this level will therefore be influenced by the personal interest of a particular political party and/or candidate.

Global warming truly is a worldwide problem; it does not care about color, class, wealth or political affirmations. There is no doubt that the scientific community has made it very clear that we are in serious trouble if we don't do something to remedy the current situation, and although it was hoped that the Kyoto Protocol would've delivered more results by now, it is definitely a step in the right direction.

- **Carbon Offsetting**

Another scheme that has been put in place in order to continue the battle with Global warming is called 'carbon offsetting', so how does it work?

Businesses can purchase 'carbon credits' in order to offset their emissions; this purchasing of credits in turn funds essential environmental projects that without this investment would be economically unfeasible. These projects play an instrumental role in the mitigation of climate change, some example of such projects are:

renewable energy; forest protection and reforestation projects.

How does a business purchase 'carbon credits'?

Carbon credits are sold in metric tons of carbon dioxide. The type of projects that sell these carbon credits are; wind farms - that displace the use of fossil fuels; household device projects - that reduce the fuel requirements for cooking and boiling water and forest protection projects - who work to prevent illegal logging. These projects and many others like them have to prove that they need 'carbon financing' through the sale of carbon credits, in order to be both financially viable and achieve reductions in greenhouse gas emissions.

Every 'carbon credit' that is generated by a project is assigned a unique identification number. When a business purchases carbon credits in order to reduce its carbon emissions, these carbon credits are 'retired' through a third-party registry. Once a carbon credit has been 'retired', it ensures that the business cannot only claim the emission reduction that is associated with it, but also that the 'retired' credit cannot be sold on again to anyone else.

In layman's terms; if you offset one ton of carbon with a carbon credit, there will be one less ton of carbon in the atmosphere than there would have otherwise been. Of course for that carbon credit to be genuine, the project must meet the essential criteria. It is compulsory that a project is able to prove that it reduced the emissions by the level it claimed, as well as any leakage issues that might have arisen. Leakage is when the reduction in emissions in one area might inadvertently increase the emissions in another.

To make sure that they are adhering to the very highest standards, projects make sure that they use third-party standards of the highest order, such as; the 'Verified Carbon Standard' (VCS), Gold Standard and Climate Action Registry (CAR), amongst others. In addition to this they must also use independent third parties to corroborate their claims.

Scientific consensus claims that the carbon emissions must be reduced by 80% by the year 2050 in order to avoid a temperature rise of more than 2 degrees Celsius. The role of purchasing 'carbon credits' provides an 'indirect' route in offsetting their carbon emissions but not a 'direct' one, however it does all in all work out pretty well, as a kind of symbiotic relationship develops between the two companies, which ultimately works to help sustain the environment.

A good carbon management program needs to include internal reductions such as a decrease in; energy use, business travel and waste etc. The reduction of emissions in these areas is however often accused of being either 'cost-prohibitive' or having a negative impact on performance. This is why a really good carbon management program is needed to ensure that emission can be reduced without compromising performance and/or profit.

Both the 'Kyoto protocol' and the process of 'Carbon offsetting', are things that we as laymen have no control over. In the next chapter, we will be looking at how you can personally impact on greenhouse gas levels by taking responsibility for, and reducing your own carbon footprint.

CHAPTER 6: THE 'SHORT' STEPS TO REDUCING CARBON FOOTPRINTS

We have already looked at the biggest culprits when it comes to greenhouse gas emission, so in this chapter we will be focusing on the steps that we need to take to minimize our 'individual' footprints.

Although any reductions we can make to our footprints may appear small, they are definitely not insignificant. It is important to remember that a collection of small reductions soon adds up to something much larger, ensuring that collectively we can become an integral part of the bigger picture.

- **Household**

As far as our household footprint is concerned, we really can do a lot of things to ensure that we keep the consumption of electricity under control.

Heating

Reduce your household thermostat to below 70 degrees in the winter and raise it above 72 degrees in summer.

Lowering your hot water temperature by as little as 20 degrees (i.e. 140 to 120 degrees) will reduce CO2 emissions by the equivalent of 200 pounds of carbon a year.

Remember; the process of heating your household equates to approximately 41% of your total energy expenditure.

<u>Appliances</u>

Ensure all your electrical appliances are unplugged when they are not in use. Did you know that cell phone chargers, game consoles etc use electricity just by being plugged in? This simple action could save an average $100, not to mention a reduction in your carbon emissions.

Make sure that your fridge and freezer are regularly defrosted, it is also important to ensure that there are a few clear inches on every side of your refrigerator so that the heat can escape. Avoid keeping your fridge open for long periods of time as this allows more cold air to escape, resulting in the fridge having to work harder. Do not put warm or hot food into the fridge as this will also force the fridge to work much harder and therefore consume more electricity.

Electrical items that are left in standby mode continue to use approximately 85 percent of the electricity that they would use if they were being used.

Consider drying your clothes outside whenever possible, electrical dryers consume vast amounts of electricity.

When using a washing machine, ensure that there is always a full load and that the temperature is turned down.

Use the microwave instead of the stove. Microwaves are not only faster but also use significantly less energy. In fact a meal that takes one hour to cook on the stove, will take fifteen minutes to cook in the microwave. If you use the microwave regularly instead of the stove you will find that in time your energy bill decreases as well.

Make sure that you choose a laptop instead of a desktop PC. Desktops are far less energy-efficient than a laptop. In fact a laptop is up to 80 percent more energy efficient than your desktop, this is because of the energy-efficient adapters, LCD screens, CPUs, battery life and hard drives.

Lighting

Replace all of your standard light bulbs and appliances with Energy-Star approved models. Regular incandescent bulbs actually waste 90 percent of the energy they consume in the form of heat.

If you replace 5 incandescent bulbs with energy efficient ones, according to EPA estimates you will save at least $60 per year.

Remember; lighting consumes up to 15 percent of the total energy used in your home.

If you currently light your garden, you should consider using solar-powered lights. These will charge during the day and light up at night, requiring no electricity at all.

Diet

Make sure that you eat more fresh fruits and vegetables than meat. This is because the production of livestock is responsible for 18% of greenhouse gas emissions.
It is also beneficial for you to choose certified organic and locally produced foods when possible as the production of these foods does not involve the burning of fossilized fuels.

Paper

We live in a world of paperless communication however we still use a lot of paper completely unnecessarily. Replace any paper towels and napkins with reusable cloths, and always ensure you are buying recycled products containing at least 30 percent postconsumer waste. Products that bear the Forest Stewardship Council Logo are also great, as this means they come from well-managed forests.

It is of course important to keep up with the news on a daily basis however why subscribe to a 'paper' version of a publication when an online version is available.

If it important for you to have a paper version; then it is essential that you ensure you are recycling it on a regular basis.

Just as importantly, if you are getting your publication via the internet using a laptop, tablet and/or e-reader, make sure that you do not leave the device plugged in unnecessarily.

Water

Ensure all plumbing fittings are water-efficient; things like faucet aerators, showerheads and low-flow toilets can ensure that you save thousands of gallons of water a year.

You can also do some really simple things that can have a major effect on your overall water consumption and therefore your emissions, for example: turn off the tap while you are brushing your teeth; when you are boiling water for your cup of tea, only boil the amount of water that you need; make sure that your hot water tank is insulated with a thick jacket; have a shower whenever possible as opposed to a bath, it makes a huge difference to the amount of water used.

Plastics

Cut back on the amount of plastic products you use. Plastics are made from petroleum which is a non-renewable resource. Reducing their use will not only cut the emissions involved in their production, but also in their disposal.

If you are still in the habit of buying packaged bottled water then start filtering your own water. The issue with bottled water is not only that the bottles are made of plastic but also the emissions expelled as a result of their transportation & disposal.

Remember; tap water in most western countries is perfectly safe to drink.

- **Travel**

Personal vehicle

As far as driving is concerned, there are plenty of things you can do to reduce your footprint by conserving fuel; drive as smoothly as possible; avoid heavy braking and acceleration; never keep the engine idling; check your revs, change up before 2500rpm (petrol) and 2000 rpm (diesel); the most efficient speed is approximately 55-65 mph; check the pressure in your tires as under-inflated tires are not only dangerous but can increase the level of fuel consumption by up to 3 percent; only use air-conditioning when absolutely necessary and always make sure that you get your car serviced on a regular basis.

All of the techniques mentioned above are great however the real reduction in emissions only comes when you don't drive at all. The thought of this shocks many people however if you stop and think, there are many times that if you're honest with yourself you really don't need to drive, for example going to a supermarket which is only a few blocks away!

Think about cycling or perhaps walking – it's a much better and healthier alternative.

Flying

Do you really have to fly? Try to consider alternatives; perhaps replace those flights with bus or train rides. Alternatively you might wish to consider the possibility of combining trips so that you minimize the total amount of time you spend in the air.

Remember; airplanes use a vast amount of fuel, particularly during the process of taking off and landing. It is therefore always better if you can take direct flights and avoid stopovers.

There are many websites out there nowadays that show you how you can travel all over the world without ever having to get on a plane; make sure you check them out before booking your flight.

Plant a tree

One of the best things you could ever do for your environment is to plant a tree as it is a brilliant way of giving back to the planet for everything it has done for us. Trees can help safeguard our environment as they consume carbon dioxide and give out oxygen, another bonus is that they also provide us with much-needed shade.

According to the 'Urban Forestry Network', a single tree absorbs 13 pounds of carbon dioxide every year and as they grow older this figures increases; in fact they can end up ultimately absorbing 48 pounds of carbon dioxide annually. In addition to this, one ten-year-old tree releases enough oxygen into the atmosphere to support two human beings!

Throughout this chapter we have highlighted numerous things that we can do to reduce our carbon footprint, and although they may appear to be are really small things that can be done without compromising our lifestyles, it is really important to remember that all these 'little things' add up, and over time can significantly reduce our collective carbon footprints.

The choice is ultimately ours. We can either decide to be more responsible by striving to create a better world for the generations to come, or we can choose to ignore all the suggestions that have been presented in this chapter and go about our lives as though they really don't make a difference.

Think about your children and future generations. You only have only one life; make it matter!

Chapter 7: Interesting Facts about Carbon Emissions

Now that we have discussed lots of different ways in which we can effectively reduce our carbon footprint, let us now educate ourselves with some interesting facts about 'carbon emissions' that you may not know about.

According to the U.S. Environment Protection Agency, the atmospheric concentrations of carbon dioxide; the main emission responsible for your carbon footprint, have increased by 36 percent since 1750. The EPA and other environmental agencies developed carbon footprint calculators to measure the carbon contributions made by individuals.

The concept of the 'carbon footprint' was formulated in 1979 by a U.S. senate energy committee discussion about the 'environmental footprint' of government operations in Yosemite National Park. The chief operating officer of Green Mountain; Tom Rawls, is largely credited with coining the term 'carbon footprint' in an article in the Seattle Times titled 'Carbon Count: Forests Enlisted in Global Warming War' which was published on November 18, 2000. The term gained even greater prominence when it was used in the British Petroleum advertising campaign in 2005.

Fossil fuels are primarily responsible for the release of carbon dioxide and other greenhouse gases into the atmosphere. They are the main source for almost 80 percent of the world's industrial energy, and are used to both heat our homes and fuel our cars. Fossil fuels are however a non-renewable resource and as a result will eventually run out.

It is important to understand that greenhouse gases are not inherently 'bad'; they help to keep the earth warm and conducive to life, and are in fact integral to the survival of humans, bird and animal. Global warming only occurs when these gases are over produced, building up in the atmosphere to such an extent that problems start to occur.

Global warming began in the early 1700's as a result of the increase in the use of fossil fuels during the Industrial Revolution. We currently have such a dependence on these fuels for our factories, homes and cars etc that it is now very difficult to imagine switching to alternative, healthier options.

The U.S. alone emits around 6 billion tons of carbon dioxide every year; 40 percent of which comes exclusively from power plants.

There is more carbon dioxide in the air today than there has been in the last 800,000 years.
If you preserve one acre of forest land in the United States, it will prevent approximately 260,000 pounds of carbon dioxide from building up in the atmosphere.

As the sea temperatures increase, its ability to retain carbon dioxide decreases, as a result higher levels of carbon dioxide are released into the atmosphere.

The ice sheets that are covering Greenland, East Antarctica and West Antarctica have already begun to melt rapidly due to increasing sea and atmospheric temperatures. It is estimated (Hansen and Sato, 2011) that the rate of the loss of these ice sheets, could double every decade.

Sea levels are also rising at an alarming rate and this will eventually result in the flooding of low-lying coastal areas. This is particularly alarming when you consider that 70 percent of the world's population lives in what are considered high-risk flood zones. The same threat hangs over many of our incredibly fertile river deltas that are currently used extensively for producing food.

For every one meter that the sea level rises, the coastline will be eroded by 100 meters. This obviously poses a very serious threat to our infrastructure as many roads, bridges, airfields and even domestic and commercial buildings are built on low-lying coastal areas.

Scientists have predicted for many years that the increase in air temperatures will also increase the amount of water vapor in the air, therefore changing the air flow and rainfall patterns in some areas. These changing weather patterns will result in an increase in the amount of rainfall in some tropical and high latitude areas, putting them at risk of serious flooding, but on the flipside will also cause other areas to suffer from serious droughts.
These changes in weather patterns will also put some species of plants and animals at very real risk of extinction; this in turn will then put those humans that are dependent on these plants and animals for food at risk.

As the amount of CO2 in the air increases, the quantity of it that is absorbed by the colder areas of seawater also increases, this excessive absorption of CO2 causes the water to become more acidic. This in turn reduces the calcites in the water that are required by some marine animals to form shells, e.g. corals. Consequently the availability of iron in the water needed by phytoplankton also reduces. The loss of these animals poses a very serious threat to the overall marine food chain upon which both humans and animals are dependent on for their food.

Conclusion

We have learnt throughout the course of this book, not only what a 'carbon footprint' is, but also how important it is for us to realize that we all contribute to the size of the cumulative carbon footprint that is currently posing such a serious threat to our world in the form of global warming.

It is recommended that we decrease our carbon footprint by at least 80 percent if we want to have any level of control over things to come in the future. If each one of us does our bit in lowering our own carbon footprint, we can make a massive difference to the worlds combined carbon footprint.

The process of global warming might appear to be happening at a very slow rate, but over the course of time it will prove to be catastrophic to the planet we live on. It is up to us to become the engineers of change and reverse the phenomenon that is slowly eating its way into the fabric of our environment; if we do not do this now, it will become increasingly difficult to save our planet from the impending doom.

We have detailed in this book the various ways in which we can reduce our carbon footprint and also how to measure it. It is now time that we take action to reduce the amount of carbon dioxide being pumped into the atmosphere and therefore reverse the adverse effects of global warming in order to secure a better future for this planet for years to come!

So what are you waiting for? Let's get started by implementing what you have learnt, and start making a real difference.

Made in the USA
Las Vegas, NV
09 November 2020